Tumbled Stones Picture Books

Volume 6: Common Stones 2

S Murphy

Purpose of this book

My hope is that this book and series will be a reference to help identify tumbled stones. Most books only show one example, or show the rough specimen, this book shows a picture of numerous stones. Also included is a very short description and where applicable, other names or similar stones. I am not an expert, just someone who really likes rocks. I find the different patterns and colors of tumble stones very interesting. The smaller close up picture is just for fun. The book is laid out so that if you are in the 2 page view, some similar stones can be compared.

Hopefully the information will help you to look further on the internet or other books to obtain the information that you want to find, for example, crystals for healing, Feng Shui, the chemical makeup of the stone, or maybe you just like to look at stones.

I personally do not care much for dyed or manmade stones, but have included a few in this book and series as they are quite common.

Tumbled Stones Picture Books

Volume 1: *Common Stones*
Volume 2: *Pinks, Rubies and More*
Volume 3: *Reds*
Volume 4: *Orange and Garnets*
Volume 5: *Yellows*
Volume 6: *Common Stones 2*
Volume 7: *Greens and Serpentine (coming soon)*

Table of Contents

Mangano Calcite

Mangano Calcite is a light pink opaque stone, often showing banding.

Other Names:
Inca Rose

Rhodochrosite

Different shades of pink occur when Rhodochrosite has bands.

Thulite

Thulite is the pink variety of zoisite.

Other Names:
Desert Jasper

Polychrome Jasper

Polychrome Jasper is very colorful with pinks, blues, browns and cream coloring.

Brecciated Jasper

Brecciated Jasper is formed when jasper breaks and then the cracks are filled in.

Other Names:
Mook Jasper

Mookaite Jasper

Mookaite Jasper is a jasper with yellow, red and purple coloring.

Sunstone

Sunstone is a feldspar with inclusions.

Garnet

There are many different varieties of garnet. The ones shown here are reddish in color.

Citron Chrysoprase

Other Names:
Lemon Chrysoprase
Yellow Magnesite
Nickeloan Magnesite

Citron Chrysoprase is actually a yellow form of magnesite.

Other Names:
Golden Quartz

Yellow Quartz

Quartz occurs in many colors, shown here is the yellow variety.

Lionskin

Lionskin is a combination of quartz and tiger's eye.

Rutilated Quartz

Rutilated Quartz is quartz with inclusions of rutile, here they are golden in color.

Fancy Jasper

Fancy Jasper consists of many different colors.

Other Names:
Infinity Stone

Infinite

Infinite is a type of serpentine with cream and lighter green coloring.

Prehnite

Prehnite is a light green stone with a pearly luster.

Chrysoprase

A variety of chalcedony, Chrysoprase can be found in many shades of green.

Chrysocolla in Quartz

Chrysocolla in Quartz is the bluish-green of chrysocolla in a whitish quartz matrix.

Other Names:
Cordierite
Water Sapphire

Iolite

A blue to bluish-gray stone, Iolite can also be yellowish when the light angle changes showing pleochroism.

Dumortierite

Dumortierite is a dark blue stone with black spots or lines.

Blue Quartz

Quartz occurs in many colors, Blue Quartz is the blue variety.

Ametrine

Ametrine is naturally occurring with bands of yellow citrine and purple amethyst.

Fluorite

Fluorite occurs in many colors, often with multiple colors in one stone.

Brown Zebra Jasper

Brown Zebra Jasper has zebra like patterns in brown and cream colors.

Other Names:
Fossilized Wood

Petrified Wood

Petrified Wood can be seen in many different colors, it is fossilized wood.

Apache Tears

Apache Tears are a form of obsidian, when held up to the light they are translucent.

Mahogany Obsidian

Another type of obsidian, Mahogany Obsidian has reddish-brown markings.

Other Names:
Turritella Agate

Turritella

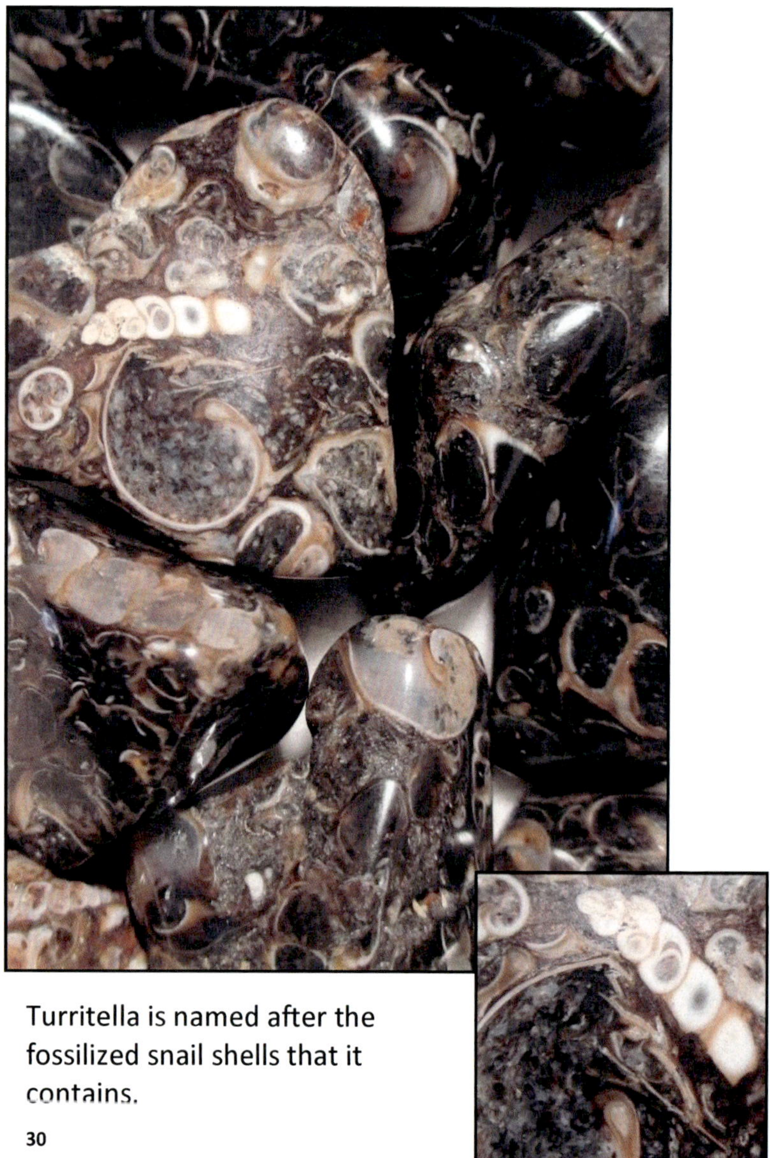

Turritella is named after the fossilized snail shells that it contains.

Other Names:
Picasso Marble
Picasso Jasper

Picasso Stone

Picasso Stone is a marble consisting of black, gray, brown and occasionally some white coloring.

Black Tourmaline

Other Names:
Schorl

Tourmaline exists in many different colors, shown above is the black variety.

Black Obsidian

Obsidian is a volcanic glass, different colors and markings exists.

Tourmalinated Quartz

Other Names:
Tourmaline Quartz

Tourmalinated Quartz is quartz with tourmaline inclusions.

Other Names:
Rock Crystal

Clear Quartz

Clear Quartz is a clear, colorless variety of quartz.

Magnesite

Normally white to cream colored, Magnesite can be found dyed in many colors.

Other Names:
Orbicular Jasper

Ocean Jasper

Ocean Jasper is an orbicular jasper with lots of different colors and circular patterns.

Crazy Lace Agate

Crazy Lace Agate is an agate with many bands and colors.

Bronzite

Bronzite is a stone with a bronze like metallic luster.

Other Names:
Aventurine Glass

Goldstone

Goldstone is manmade by mixing copper and glass.

Other Names:
Sea Opal Glass

Opalite

Manmade, Opalite is glass that has lots of iridescent color.

Index

Resources, References and Places to Buy:

Bonewitz, Ronald Louis. *Rock and Gem.* New York: DK Publishing, 2005. Print.

fortheloveofcrystals.com

Hall, Cally. *Gemstones, Eyewitness Handbooks.* New York: DK Publishing, 1994. Print.

ksccrystals.com

Lilly, Simon and Sue. *The Essential Crystal Handbook*. London: Duncan Baird Publishing Ltd., 2006. Print.

www.metaphysicalrealm1.com

www.Mindat.org

www.Mineral.net

www.ravenandcrone.com

The Rock Room on eBay

www.rocktumbler.com

Soulfulcrystals.co.uk

Wikipedia.com

Made in the USA
Middletown, DE
09 January 2018